Ephraim Cutter

Partial Syllabic Lists of the Clinical Morphologies

of the blood, sputum, feces, skin, urine, vomitus, foods, including potable waters, ice and the air, and the clothing

Ephraim Cutter

Partial Syllabic Lists of the Clinical Morphologies
of the blood, sputum, feces, skin, urine, vomitus, foods, including potable waters, ice and the air, and the clothing

ISBN/EAN: 9783337390297

Printed in Europe, USA, Canada, Australia, Japan

Cover: Foto ©berggeist007 / pixelio.de

More available books at **www.hansebooks.com**

PARTIAL SYLLABIC LISTS

OF THE

CLINICAL MORPHOLOGIES

OF

The Blood, Sputum, Feces, Skin, Urine, Vomitus, Foods, including Potable Waters, Ice and the Air, and the Clothing (After Salisbury).

BY

EPHRAIM CUTTER,

M.D. Harvard and University of Pennsylvania, A.M. Yale, LL.D. Iowa, Hon. F.S.Sc. (London).

Principal Medical Department, American Institute of Micrology; First to Photograph Consumptive Blood; Inventor Several Forms of the Clinical Microscope, The Cam Fine Adjustment, etc., etc.

Corresponding Member Société Belge de Microscopie and Gynecological Society of Boston; Associate Member Philosophical Society of Great Britain; Honorary Member California State Medical Society; Member American Society of Microscopists, American Medical Association, etc.

Author Boylston Prize Essay, 1857; Primer of the Clinical Microscope; What I Use the Microscope For; Morphology of Diseased Blood; Morphology of Rheumatic Blood (Ninth International Medical Congress); Morphology of Potatoes, Cooked; Crypta Syphilitica; Monstrous Spermatozoa; Micrographical Contribution as to the Vegetable Nature of Croup; Tubercle Parasite; Microscopical Examination of Ice; Suspicious Organisms in the Croton; Beri-Beri; Trichina; Butter; Effects of Alcohol on Brain Tissues; Action of Alcohol on the Blood; Asthmatos Ciliaris; Diphtheria and Potatoes; Use of Microscope in Consumption; Throat Syphilis and Tubercle according to Salisbury; Tolles' 1.75 inch Objective, its History, Use, and Construction; Amœboid Movements of the White Blood-Corpuscle; A New Sign of the Pre-Embolic State; Food Stuffs under the Microscope, etc., etc.

"A capacity to do good not only gives a title to it, but makes the doing of it a duty."—*Duke of Brandenburg*, 1691.

NEW YORK
THE ARISTON, BROADWAY AND 55TH STREET
PUBLISHED BY THE AUTHOR
1888

COPYRIGHT BY
EPHRAIM CUTTER,
1888

Dedication.

This work is respectfully dedicated to the following, who have shown themselves searchers after medical truth and courteous to co-laborers.

Benjamin Cutter, M.D., A.M., in memoriam, summa laude
J. Marion Sims, M.D., LL.D., in memoriam, summa laude
E. S. Gaillard, M.D., LL.D., in memoriam, summa laude
Louis Elsberg, M.D., in memoriam, summa laude
George Waterhouse Garland, M.D., in memoriam
George M. Beard, M.D., in memoriam, summa laude
S. D. Gross, M.D., LL.D., D.C.L., in memoriam, summa laude
Frank H. Hamilton, M.D., LL.D., in memoriam, summa laude
James R. Nichols, M.D., in memoriam, summa laude
Washington L. Atlee, M.D., in memoriam, summa laude
Professor L. A. Sayre, M.D.
Professor T. G. Thomas, M.D.
Professor Albert Vander Veer, M.D., Ph.D.
Professor R. J. Nunn, M.D.
Professor T. E. Murrill, M.D.
Professor T. E. Satterthwaite, M.D.
Professor Joseph Jones, M.D.
Professor Jacob Cooper, M.D., Ph.D., J.C.D., S.T.D.
Professor Wm. B. Atkinson, M.D., A.M.
Professor Byron Stanton, M.D.
Professor J. Solis Cohen, M.D.
Professor W. W. Dawson, M.D.
Professor Samuel B. Ward, M.D., Ph.D.
Professor Joseph Leidy, M.D.
Professor James P. Boyd, M.D., M.A.
Professor D. Hayes Agnew, M.D., LL.D.
Professor D. Humphreys Storer, M.D., LL.D.
Professor H. M. Field, M D
Eugene Van Slyke, M.D.
Israel H. Taylor, M.D.
George D. Dowkontt, M.D.
Ezra P. Allen, M.D., Ph.D.
David Prince, M.D.
Alfred C. Garratt, M.D.
G. L. Simmons, M.D.
W. Symington Brown, M.D.
Jonas C. Harris, M.D.
Austin W. Thompson, M.D.
Samuel W. Abbott, M.D., M.A.
J. J. Mulheron, M.D.
R. E. Thompson, M.D., F.R.C.P. Lond., summa laude
Henry O. Marcy, M.D., LL.D.
J. N. Hyde, M.D.
Landon B. Edwards, M.D.
Sir James Grant
Professor Aust-Lawrence, M.D.
D. H. Goodwillie, M.D.
Professor A. B. Arnold, M.D.
R. U. Piper, M.D.
W. R. Weisager, M.D.
Professor Domingos Freire, M.D.
Caleb Green, M.D.
A. F. Pattee, M.D.
Fr. Ecklund, M.D.
Professor E. A. Wood, M.D.
Professor M. C. White, M.D.
M. G. Wheeler, M.D.
Henry C. Bunce, M.D.
Sir Morell Mackenzie

AND

DEDICATION.

TO MY INSTRUCTORS

JAMES H. SALISBURY, M.D., LL.D., maxima laude
PROFESSOR PAULUS F. REINSCH
PROFESSOR J. P. COOKE, M.D.

PROFESSOR OLIVER WENDELL HOLMES, M.D., LL.D., D.C.L.
REV. JOSEPH COOK
GEORGE B. HARRIMAN, D.D.S.

ROBERT B. TOLLES, in memoriam

INTRODUCTION TO THE MOR-
PHOLOGIES.

It is now over ten years since the writer first applied this word to the account (logos) of the forms (morphos) found in the blood, sputum, fæces, urine, etc., and its general adaptation seems to justify the use of the term. It was employed to facilitate the introduction of the thoughts and results embraced in The Relation of Alimentation to Disease* by J. H. Salisbury, M.D., LL.D., the master discoverer and explorer.

The morphologies of his discoveries are over twenty-five years old. The number of people who have been cured by the thorough and systemic plans based on them is such that there is no need of apologizing for bringing them more prominently to notice, but rather of apologizing that they have been kept back so long. The writer has not ceased night and day to urge their publication, and he is permitted to hint gently that, if what has now been issued is well received, much more valuable

---
* New York: J. H. Vail & Co., 1888.

treasures will be dispensed from the storehouse to all who ask for them.

Those who gain a tolerable knowledge of these lists will expect, among other things, to diagnosticate consumption of the lungs in (1) The pretubercular state; (2) In the invasion stage; (3) In the breaking down stage. To diagnosticate syphilis at once. To diagnosticate rheumatism, in its various forms. To diagnosticate fibræmia, anæmia, leucocythæmia, malaria, diseases of fatty degeneration, sclerosis, locomotor ataxy, impending apoplexy, and paraplegia, etc., etc. To diagnosticate a state of perfect health, a tendency to diseased conditions, etc., etc.

Since nearly a quarter of a century has been spent on these morphologies, it cannot be claimed that they are hastily gotten up; still the lists are all partial, subject to addition and subtraction, as need requires. They may be taken to represent the actual state of knowledge at the present day, which is quite an advance over a quarter of a century ago.

PREFACE.

For some years the writer has needed a published list of the clinical morphologies for the use of his pupils. He has waited long to have the lists complete, but in vain. Complete knowledge of any subject is about as rare as a completed city. Knowledge is ever on the increase, like most of our cities. We use our cities even if incomplete, so must we use our knowledge as far as it goes.

One object of this work is to show the height and depth, the length and breadth of the so-called Salisbury plans; that they are entitled to respectful hearing; that they include a wide survey and comprehensive grasp of the world that comes in contact with our bodies, outside and in; that they have no narrowness of range nor contraction of vision; that they deal with facts more than with opinions; that the tests to which they may be subject are close at hand and near to reach. *They are cis- not transatlantic.*

These morphologies also show that the writer has not ridden a one-horse hobby in satisfying his mind of the truth of the plans named, but that he has endeavored to take broad views of all the evidence in the matter before coming to conclusions.

It has been suggested that the writer give a short history of his relation to these subjects. In justice to all concerned, and to make shorter work, the personal pronoun will be used, mostly.

I began the use of the microscope as a means of education and useful knowledge, if my memory serves me rightly, in the Sheffield Scientific School of Yale College in 1850. The winter of 1853-4 I spent in Professor J. P. Cooke's private laboratory, working up the morphologies of blood and urine, together with their micro-chemistry. Besides him, I have studied under Dr. O. W. Holmes, Col. J. J. Woodward, G. B. Harriman, D.D.S., Professor Paulus F. Reinsch, the highest authority in algæ; and Dr. James H. Salisbury. The last gentleman excels all others in the amount of original information which I have found of priceless need and value in medicine. Before studying medicine, I was inspired with a desire to know all I could about the causes of disease. Having, from a child, been in the habit of accompanying, in his professional rounds, my father, the late Dr. Benjamin Cutter, of Woburn, Mass. (who honored his profession for forty years), I early took in the idea that there was a great field of much-needed effort, from the chance remarks he would drop when he resumed his seat in the carriage (which I kept) after having seen some very sick patients. He said often, "Oh, how I wish we doctors knew more as to the real causes of disease." This impulse was much strengthened by his telling me (when I informed him that I did not want to study medicine to practice it, but only to know the causes of disease), "Go ahead; study all you can. I will help all I can, but I want you to study these three things.

"1. What is the cause of consumption.

"2. What is the cause of the diseases of women.

"3. What is the cause of diseases of the nervous system.

"We doctors do not know anything about them." And yet this was a surgeon who successfully, without anæsthesia, opened the knee joint and removed a free cartilage (assisted only by the writer when twelve years old).

The present work is the outcome of this paternal injunction. Advisedly, seriously, and thoughtfully can it be now said, these three (3) problems have been answered satisfactorily, and we know that unhealthy alimentation causes primarily all of these classes of disease.

In 1857, the Boylston Prize was awarded the writer for an essay on "Under what Circumstances do the Usual Signs Furnished by

Auscultation and Percussion Prove Fallacious?"

In 1858, the writer invented a laryngoscope, which was made by Alvan Clarke & Son, the great telescope makers.

In 1866, the writer took the first photograph of the vocal cords (his own), which showed the thyroid insertion.

In 1866, he demonstrated to large numbers his own larynx in situ naturali, and the posterior nares, showing either Eustachian tubes at will, the vomer, and turbinated bones, and first demonstrated the erection of the mucous membrane of the turbinated bones in smelling odorous or malodorous substances.

Before 1860, he travelled over five thousand miles to see if alcoholism could be connected with consumption of the lungs.

In 1867, I visited Dr. Salisbury to learn how to study malaria. At that time, I found he had gone one step farther than I, and connected the vinegar plant with consumption. Thus he supplied the missing link to my chain, and, after repeated and careful observations, I came to learn the truth of this new doctrine in the actual treatment and cure of cases, and ever since have endeavored to make it known in proper ways, so far as I could.

Finding Dr. S.'s drawings denounced and ridiculed, and, of course, rejected, and stung to think that this work should be deemed an idle tale, I set myself to work to photograph as many of the appearances in consumptive blood as I could. Probably this was the first attempt of this kind. Never before this had I known of any blood being photographed save for medico-legal purposes. I found the subject very much hampered with details which I thought should be done away with. Feeling the greatness of the work, and that it should be done before my eyesight and faculties were too old, I gave up a fine country practice and settled in Cambridge, Mass., as I thought this seat of learning would be the most favorable for the encouragement and prosecution of my work. The winter of 1875–6 was spent in working up micro-photography. Fortunately I came across Dr. G. B. Harriman, Surgeon-dentist, of Boston, who possessed magnificent objectives made by R. B. Tolles, among them the 1-50 inch and 1-75 inch. He entered into the work heartily, and together we took micro-photographs of consumptive blood morphology for the first time and with the highest powers ever used up to that time and since (so far as I can ascertain), and which have been pronounced good in Europe.

The account of this work may be found in the *American Journal of Science*, New Haven, August, 1879; *Scientific American Supplement*, September, 1879; *Journal of Micrographie*, Paris, 1879. These photographs have been placed on the screen before the American

Medical Association; Chicago Medical Society; Academy of Medicine, Virginia; Academy of Sciences, New York; Albany Medical College; Monday Lectureship, Boston, Rev. Joseph Cook; Gynecological Society of Boston, and many other bodies. These things are named to show that I am in earnest, for none would have done this unless he was sincere and meant what he said.

In 1876, Professor Paulus F. Reinsch was introduced to me at the Botanical Garden, in Cambridge, as the greatest algologist. Careful study with him has confirmed my views on these so-called Salisbury plans. So many cures have followed, that I feel it would be a crime in me not to testify to what I know, and how I have been set right upon the three tasks propounded by my honored and honorable father more than thirty years ago, and which, so far as in me lies, I have tried to solve or have solved. I am a co-witness with Dr. Salisbury; "that in the mouth of two or three witnesses every word may be established." I charge therefore those to whom these presents may come to look over the evidence, and take time before they treat these things as "idle tales." If the "Imperial Granum" which I have shown morphologically to be common flour, and which the Connecticut agricultural experiment station has also shown to be common flour, selling at $1.00 per pound, while it is worth from

$0.025 to $0.05, is used and indorsed by the medical profession (so that its proprietors have become rich and use fifty-two barrels of flour in one batch), on statements that wilt before the microscope and crucible, does it look well for the same noble profession to treat the plans here indorsed, which stand the tests of the microscope and chemistry, as an "idle tale?"

I have nothing but good feeling or words towards those who honestly differ, but I do dislike to see physicians led by persons who not only have no medical education, but also advertise untruths and at the same time consider these plans as "idle tales," and neglect to look into the evidence which has stood for nearly a quarter of a century, and which affects the weal or woe, not only of the public, but of the profession and the very gentlemen themselves.

Be this as it may, in time to come, none can accuse me of not having tried to discharge the duties which every physician owes to his fellows, to wit: if any physician knows or thinks he knows anything which will better the practice of medicine, there is a moral obligation for him to discharge by making it known, and so long as the rules of courtesy are observed by the contributor, he is entitled to a courteous hearing. Any departure from this savors of savagery and puts the doer at once out of the pale of civilized ethics.

MAY 1ST, 1888.

CONTENTS.

	PAGE
Dedication,	iii.–iv.
Introduction,	v.–vi.
Preface,	vii.–xiii.

I. The Morphology of the Blood—Mode of Study, . 1
  A. General list of the Morphology of the Blood in Health and Disease, 3; The Colored Corpuscles, 3; The Colorless Corpuscles 4; The Serum, . 4
  B. Morphology of the Blood in Health, . . 7
  C. Movements and Changes of the Blood in Dying, . 9
  D. Morphology of the Blood in Consumption of the Lungs ; Use, 9 ; First or Incubative Stage 10 ; Second Stage, of Transmission, 10; the Third Stage or Stage of Tubercular Deposition 12 ; Fourth Stage, Interstitial Death, . . . 12
  E. The Morphology of the Blood in Rheumatism, . 13
  F. Fibræmia, ... 17
  G. Thrombosis, ... 17
  H. Embolism, . . 18
  I. Pre-embolic State, . . . 18
  J. Anæmia, ... 18
  K. Pernicious Anæmia, ... 19
  L. Morphology of the Blood in Syphilis, . . 19
  M. Morphology of the Blood in Eczema, . . 20
  N. Morphology of the Blood in Scrofula, . 21
  O. Morphology of the Blood in Malaria, . . 21
  P. Hereditary Taints, ... 22
  Q. Cancer, ... 23
  R. Morphology of the Blood in Variola and Vaccinia, 23
  S. Morphology of the Blood in Typhoid Fever, . 23
  T. Morphology of the Blood in Scarlet Fever and Diphtheria, ... 24
  U. Morphology of the Blood in Fatty Degeneration, . 24
  V. Morphology of the Blood in Fibrous Consumption, 25

	PAGE
W. Cholesteræmia,	25
X. Morphology of the Blood in Carbuncle,	26
Y. Morphology of the Blood in Yellow Fever,	26
Z. Leucocythæmia,	26
II. Morphology of the Sputum,	27
III. Morphology of the Feces,	33
IV. Morphology of the Skin,	38
V. Morphology of the Urine,	43
VI. Morphology of the Vomitus,	48
VII. Morphology of Foods,	49
A. Waters of Lakes, Ponds and Water sheds; Hydrant Waters, 49; List, 50; Appendix,	54
B. Waters of Springs and Wells unconnected with Lakes or Ponds,	58
C. Ice, 63; List, 64; Appendix,	65
D. Air, 70; List,	72
E. Morphology of Foods; Animal and Vegetable, 74; Vegetable: Uncooked, 74; Cooked, 75; in the Feces, 75, 76, 77. Beefsteak: Uncooked, 76; Cooked, 76; in the Feces, 76; Adulteration, 77; Infants' Foods,	78
VIII. Morphology of Clothing,	80

EXPLANATORY.

Though, as stated in the title, these partial syllabic lists are after Salisbury, I wish to emphasize that those who read this book should, in order to get more information on the subjects noted, especially the blood, sputum, feces, urine, and skin, consult the works of Dr. Salisbury here named:

1. "The Relation of Alimentation to Disease," octavo, pp. xi., 334, plates 19. New York, 1888: J. H. Vail & Co. (See "Clinical Morphologies," consumption of the lungs. pp. 9 to 13; fibræmia, p. 17; anæmia, p. 18; pernicious anæmia, p. 19; fibrous consumption, p. 25; sputum, pp. 27 to 32; feces, pp. 33 to 37.)

2. "Microscopic Examinations of the Blood and Vegetations Found in Variola, Vaccine, and Typhoid Fever." 66 pages and 62 illustrations. New York, 1868. (See page 23, " Clin. Morphologies.")

3. "Remarks on the Structure, Functions, and Classification of the Parent Gland Cells, with Microscopic Investigations Relative to the Causes of the Several Varieties of Rheumatism and Directions for their Treatment." 1 plate of illustrations. *American Journal Medical Sciences*, October, 1867, p. 19. (See pp. 13, 14, 15, 16, 17, "Clin. Morphologies.")

4. "Vegetations Found in the Blood of Patients Suffering from Erysipelas." Hallier, *Zeitschrift für Parasitenkunde*, 1873, 8 illustrations.

5. "Infusorial Catarrh and Asthma." 18 illustrations, do., 1873.

6. "Description of Two New Algoid Vegetations, One of which Appears to be the Specific Cause of Syphilis, and the Other of Gonorrhœa." Do., 1873. Also *Amer. Jour. Med. Sci.*, 1867. (See pp. 19-20, "Clin. Morphologies.")

7. "Chronic Diarrhœa and its Complications, or the Diseases Arising in Armies from a too Exclusive Use of Amylaceous Food, with Other Interesting Matter Relating to the Diet and Treatment of these Abnormal Conditions, and a New Army Ration Proposed with which this Large Class of Diseases may be Avoided." The Ohio Surgeon-General's Report for 1864.

8. "Probable Source of the Steatozoon Folliculorum." St. Louis *Medical Reporter*, January, 1869.

9. "Something about Cryptogams, Fermentation, and Disease." Do., February, 1879.

10. "Investigations, Chemical and Microscopical, Resulting in what Appears to be the Discovery of a new Function of the Spleen and Mesenteric and Lymphatic Glands." Do., August, 1867, 29 pages.

11. "Discovery of Cholesterin and Serolin as Secretions in Health of the Salivary, Tear, Mammary, and Sudorific Glands; of the Testis and Ovary; of the Kidneys in Hepatic Derangements; of Mucous Membranes when Congested and Inflamed, and the Fluids of Ascites and that of Spina Bifida." *Amer. Jour. Med. Sci.*, April, 1863, 2 plates, 17 pages.

12. "Remarks on Fungi, with an Account of Experiments Showing the Influence of the Fungi of Wheat and Rye Straw on the Human System, and Some Observations which Point to Them as the Probable Source of Camp Measles, and Perhaps of Measles Generally." Do., July, 1862, 1 plate, 30 pages.

13. "Inoculating the Human System with Straw Fungi to Protect It Against the Contagion of Measles, with Some Additional Observations Relating to the Influence of Fungoid Growths in Producing Disease, and in the Fermentation and Putrefaction of Organic Bodies." Do., October, 1862, 8 pages.

14. "Two Interesting Parasitic Diseases, One We Take from Sucking Kittens and the Other from Sucking Puppies. Trichosis Felinus and T. Caninus." *Boston Medical and Surgical Journal*, June 4th, 1868. 6 illustrations. Also *Zeitschrift für Parasitenkunde*, Hallier, Jena, 1875.

15. "Malaria," McNaughton prize essay, 1882. Octavo. pp. 152, plates 10. New York: W. A. Kellogg, 1885. (See pp. 21, 22, "Clin. Morphologies.")

16. "Diphtheria, Its Cause and Treatment." G. A. Davis, Detroit. 3 plates, 1884. (See page 24, "Clin. Morphologies.")

Which are a partial list of his works.

THE MORPHOLOGY OF THE BLOOD.

MODE OF STUDY.

It is necessary to have the patient, the microscope, the light, the means of withdrawal of the blood—a lancet, spring lancet, the scarificator of the writer, or a needle, which is not the best thing—all together.

There is no such thing as taking the blood home to examine. The changes are so rapid that most of the important ones disappear in ten minutes' time. Still, after these are gone, many valuable points remain to be looked for.

Kind of blood.—The capillary—not the venous or arterial.

Site of withdrawal.—On the radial or ulnar side of the forearm near the wrist. The skin should be clean and free from hair. If dirty, wash with soap suds or ammonia water. (It is well that the beginners should study the skin surface, dirt, and epithelium, before looking at the blood.) Take the patient's forearm in the hand, and make the skin tense in the interval between the thumb and forefinger. A quick puncture is then made, about one-eighth of an inch deep. The tension of the grip will squeeze out a drop of blood. The size of the drop should bear a direct relation to the size of the cover. If there is too little blood, the corpuscles will become crenated, that is, wrinkled from a sort of protoplasmic action induced by too much dryness in the space about the blood. If there is too much blood, the superfluity will float the cover about; there will be too much thickness of the film, and it will crowd the red corpuscles so much as to render them indistinguishable. The excess must be removed by a bibulant. Very much depends on handling the drop of blood rightly. When the drop evenly diffuses itself, it is presumed that the film is about uniform in thickness, so that one can judge somewhat as to the comparative number of corpuscles in each specimen. The process of transferring the blood should take only a few seconds of time; a fraction should be sufficient.

Of course, the slide and cover should be previously cleaned, and also the microscope should be free from dirt and in focus; as, after a previous use, if the blood specimen is placed on the stage, it will be in focus at once, and the rapid movements, changes, and morphological elements will be visible immediately.

The novice had better scrutinize carefully

everything he sees, not caring whether he knows the name of the object or not.

A. GENERAL LIST OF THE MORPHOLOGY OF THE BLOOD IN HEALTH AND DISEASE.

The Color of the Blood to the Unaided Eye. Consistence of the Blood. Rapidity of Clotting.

1. The colored corpuscles.
2. The colorless corpuscles.
3. The serum.

1. *The Colored Corpuscles.*

In normal proportion.
In excess.
In diminished quantity.
Normal consistence.
Too soft, plastic, and sticky; adhering together and being drawn out in thread-like prolongations.
Nummulated, like rolls of coin.
Not nummulated.
Evenly and loosely scattered over the field.
Slightly grouped.
In irregular, compact masses.
In ridges.
Color, clear, fresh, bright, ruddy, clean cut.
Color, pale, muddy, ashy, unlustrous, not fresh, not bright, not ruddy.

Holding firmly the coloring matter, yet soft and plastic.

High colored, smooth and even in outline, hard and rigid.

Allowing the coloring matter to escape freely, obscuring their outlines.

Mammillated.

Cholesterine in.

2. *The Colorless Corpuscles.*

In normal proportion.

In too small quantity.

In excess.

Normal in quantity or in excess; sticky and plastic, endangering the formation of thrombi and emboli.

Ragged and broken down.

In excess, ragged and broken.

In excess, smooth and even.

Containing vacuoles.

Containing vegetations that distend them to an enormous size.

Contain thin, bladder-like, empty cells, of various sizes, that distend them.

Contain the spores of crypta syphilitica.

3. *The Serum.*

Too little.

Too much.

Normal.
Its fibrin:
    In normal proportion.
    In too small proportion.
    In too large proportion.
    Meshes normal in size and in arrangement, allowing the free circulation of blood-cells through them.
    Meshes too small to admit of the free circulation of blood-cells through them, on account of which the blood-cells arrange themselves in ropy rows, or ridges and masses, being held in the meshes of the partially clotted or contracted fibrin. In such cases, the individual fibrin filaments have an increased diameter and opacity.
    Want of, in pernicious anæmia.
    Enlarged, thickened, and more opaque in rheumatism.
    Thrombi of, filled or not with granular or crystalline matters.
    Sticky and plastic.
Minute grains and ragged masses of black, blue, brown, or yellow pigment.
Fat, globules and masses of.
Amyloid matters.
Broken-down parent cells.
Thrombi of algæ spores.
Thrombi of algæ filaments.

Algæ filaments and spores without aggregation.
Fungi spores.
Fungi filaments.
Zymotosis regularis spores.
Zymotosis regularis mycelial filaments.
Entophyticus hæmaticus spores and filaments.
Penicillium quadrifidum spores and mycelial filaments.
Penicillium botrytis infestans.
Crypta syphilitica spores and filaments.
Mycoderma aceti spores and filaments.
Saccharomyces cerevisiæ.
Alcohol and acid yeasts.
Microsporon furfur.
Gemiasma, alba, plumba, rubra.
Mucor malignans.
Biolysis typhoides.
Crypta carbunculata.
Ios variolosa vacciola.
Ios vacciola.
Cryptococcus Xanthogenicus (Freire).
Cystine, granules and crystals.
Phosphates, granules and crystals.
Stelline, granules and crystals.
Stellurine, granules and crystals.
Granules and crystals of a miscellaneous character.
Conchoidine.

Pigmentine, black, brown, bronze, aniline blue, red, yellow, etc.
Cholesterin.
Leucin.
Creatin.
Uric acid and urates.
Carbonate of lime.
Inosite.

B. MORPHOLOGY OF THE BLOOD IN HEALTH.

According to Conventional Nomenclature to Aid in Studies.

Blood from Capillaries:

Color; bright, fresh, clear, ruddy, strong.
Clotting; rapid and firm.
Red corpuscles—arrange themselves in nummulations, or are scattered evenly over the field. Normal in size. Non-adhesive. Central depression well marked on both sides; periphery well rounded, clean cut. Hold coloring matter firmly. Pass readily to and fro through the fibrin filaments. Appear fresh and fair, giving an appearance of health, like a rosy-cheeked maiden full of life.
White corpuscles—normal in size. Not enlarged by internal collections of foreign bodies. Amœboid movements strong or not. Propor-

tion, one to three hundred of red corpuscles. Consistence good. Not sticky. Color a clean white. Freely moving at will.

Serum—clear and free at first sight from any form. After five minutes, most delicate semi-transparent fibrin filaments appear, forming a very light network in the field, which offers no obstacle to the passage of the corpuscles.

There should be no spores nor vegetations in healthy serum, though they may be found by very minute examination, or by letting the blood stand for several days in closely stopped phials at a temperature of from 60–75° Fahrenheit. This is not saying that spores and filaments cannot be found in blood of persons calling themselves healthy—for some diseases exist in a latent condition, like rheumatism, syphilis, cystinæmia, and consumption. I have met with people who, on finding vegetations in their blood, have decided not to accept the evidence because they deemed themselves healthy. Again, it is difficult to find a perfectly healthy person in the community; this was made public during the "late unpleasantness," when drafts were made for soldiers. The blood evidence must be taken in connection with that of the other physical signs.

THE MORPHOLOGY OF THE BLOOD. 9

C. MOVEMENTS AND CHANGES OF THE BLOOD IN DYING.

These are important and need study. They are like the behavior and manners of people that convey ideas, as they are to be gained in no other way. After one has learnt these movements in health, he will appreciate them in disease. Again, as Dr. Salisbury remarks, there are tendencies to diseased states in the blood which need detection, as they are much easier remedied than when confirmed. It is impossible to convey these ideas on paper or in drawings; they must be learned from actual observation. *The morphology of healthy blood is a most rigid test, and in delicacy and far reaching goes beyond any of the other physical signs.* When generally known and appreciated, it will be of great benefit, specially in life insurance examinations, army or navy examinations, and in the study of the best modes of physical culture.

D. MORPHOLOGY OF THE BLOOD IN CONSUMPTION OF THE LUNGS.

Use.—In diagnosis, exceeding in value auscultation and percussion, because it detects consumption of the lungs before there is any lesion of them. To show the real progress of the case by the substitution of the morphology of health more or less, to show when patients have lapsed in the treatment by eating forbidden food, and to show when there is a real cure. To repeat, most valuable of all to make out a diagnosis of consumption with as much certainty as it is possible in human affairs, and by removing the uncertainty, sometimes dreadful, of the diagnosis that accompanies the conventional first stages of consumption of the lungs.

This value is so great that it is more than a warrant for this publication to be made. It is hardly possible to overestimate the importance of this department of physical exploration.

First or Incubative Stage.

Red blood-corpuscles are less in number, ropy, and sticky, more or less, but not much changed otherwise.

Second Stage, of Transmission.

1. *Red corpuscles.*—Color pale, non-lustrous; not clear cut, not ruddy. Consistence, sticky, adhesive. Coating of neurine removed. Not so numerous as in normal blood. Owing to the increased size and strength of the fibrin and the stickiness, they form in ridges, rows, but not so marked as in rheumatic blood. They accumulate in aggregations of confused

masses, like droves of frightened sheep. They adhere to each other, and are rotten, as it were, in texture.

2. *White corpuscles.*—Enlarged and distended by the mycoderma aceti, or spores of vinegar yeast, that are transmitted into the blood stream from the intestines.

3. *Serum.*—More or less filled with the spores of mycoderma aceti or vinegar yeast. These occur either singly or in masses of spores, which is the common form in which they are found, wherever vinegar is produced.

The fibrin filaments are larger, stronger, more massive than in health, and form under the microscope a thick network which is larger, stronger, and more marked in direct proportion to the severity of the disease or the amount of accumulation.

Besides, the serum is apt to be of a dirty ash color.

The sticky white corpuscles, the massive fibrin filaments in skeins, and the yeast spores alone or combined, form aggregations, masses, collects, thrombi and emboli which block up the blood-vessels of the lungs soonest, because exposed to cold air, the most of any viscus; *the blood-vessels contract, and thus arrest the thrombi and form a heterologous deposit, which is called tubercle.*

The Third Stage, or Stage of Tubercular Deposit.

These deposits increase so long as vitality subsists in the tubercle and surroundings. When vitality ceases, the tubercle softens or breaks down. Sometimes, if the process is very slow and life slightly inheres in it, the proximate tissue undergoes fatty infiltration, which preserves it from readily breaking down.

The morphology of the blood is the same for the second and third stages of consumption.

Fourth Stage.

Interstitial Death.

Morphology of the blood in this stage is the same as in the second and third, save that it becomes more impoverished.

The red corpuscles are thinner, paler, much lessened in number, increased in adhesiveness, stickiness, and poverty. Devoid more or less of neurine.

The white corpuscles are fewer in number, more enlarged; often ragged and rough. Distended with spores of mycoderma aceti, more adhesive, and sticky.

The serum.—Fibrin filaments are thickened, stronger, more massive, and more skeins of them present. The collects of mycoderma aceti are very much larger and more numerous; in moribund cases, I have seen them so large as almost to fill the field of the microscope. They present anfractuous edges and amœboid prolongations, giving them a weird, bizarre aspect which, under the circumstances, have a portentous aspect, for the larger and more numerous the spore collects of mycodermi aceti are, the more dangerous the case.

One great proof of the so-called Salisbury plans is, that they will entirely change the morphology of consumptive blood to that of health, and the whole process can be watched and studied to the delight of all concerned.

E. MORPHOLOGY OF THE BLOOD IN RHEUMATISM.

Rheumatism may be called the

Gravel of the Blood.

Color varies from that of health to the paleness of anæmia.

Consistency and *rapidity* of *clotting* increased.

1. *Red corpuscles.*—Color usually impaired, not always; coloring matter not so firmly held as in health.

Adhesive, sticky, often drawn out into elongated lozenge-shaped bodies with pointed ends, and sometimes filamentous joining with one or more of their fellows.

Clot in winrows, ridges, and huddled masses; sometimes quite formless. This is caused by the massive fibrin filaments holding them fast, as it were, in their firm meshes. The same thing is seen in consumptive blood, but to a less degree.

2. *White corpuscles* usually enlarged; adhesive, sticking to each other and to the red corpuscles, and matters found in the serum. Indeed, it seems to be the office of the white corpuscles so far as possible to swallow and envelop any foreign substance that may find its way into the blood. Thus we find crystalline matters in the white blood-corpuscles in rheumatism, though not always.

They undergo amœboid movements as in healthy blood—they have independent locomotion. Disease does not seem to impair their automatic movements.

Often they are increased in number. If there is fatty degeneration going on, they will be found to contain fat in globules.

3. *The serum.*

Fibrin filaments—in massive, strong and sticky threads, in abundance—in meshes, which are finer than in health, visible plainly—strong and hold the red corpuscles like prisoners—in skeins, like tangled skeins of silk—in masses forming thrombi which, when fastened, form emboli.

These thrombi are apt to involve and embrace white and red corpuscles and crystalline bodies to be named below. Sometimes the fibrin filaments are found in large round strings, curled up fancifully by the motion of the blood stream, and looking like the mycelial filaments of vegetations, from which they can be distinguished by an absence of entire cylindrical outline—ragged broken edges here and there and dichotomous and polychotomous divisions of the trunk, different from vegetations of syphilis for example. It is the presence of these fibrin filaments that makes the blood ropy, adhesive, and sticky. They have the tendency to block up the blood stream and besides to be locally deposited in the tissues, specially when the circulation is sluggish, as near the extremities and the joints.

Crystalline bodies, or gravel of the blood.

These are numerous and readily recognized; some of them are as follows:—
 1. Uric acid and urates of soda.
 2. Phosphates—specially the triple phosphates of lime and soda.
 3. Oxalate of lime.
 4. Cystine. This is quite common and easily detected.
 5. Carbonate of lime, rare.
 6. Stelline and stellurine. These occur mostly in granular form in the serum, but in

old cases, where the system is saturated, they are crystalline.

7. Black, brown, aniline blue, bronze, orange, red and yellow pigments in the form of flakes or small masses are common in rheumatic blood, and may be termed gravelly matters, that should have been eliminated by the kidneys or bowels or skin.

Latent Condition of the Characteristics of Rheumatic Blood.

The morphology of rheumatic blood exists in a latent condition in persons apparently well; but when they are exposed to cold, the blood-vessels contract, catch and detain these abnormal elements, and we have a stasis of the blood which may be active or passive and manifests itself in heat, fever, pain, swelling, inflammation or passive congestion, effusion, etc., and which make up what is known as an "attack of rheumatism." The fever may result from the effects of nature to get rid of the intruders, just as a householder will become hot in expelling from his premises a thief who is difficult to get rid of. Or to use another simile, the attack of rheumatism is like the explosion of a gun. The charge in the gun is the morphology of rheumatic blood, and the cold is the pulling of the trigger. The charge may be latent in the gun for years, but it is there with its potential energy ready to become actual from an exciting cause.

F. FIBRÆMIA.

In a nomenclature which was made before the present advance of knowledge, there is difficulty in making it fit to the new era. I shall not attempt to relieve this difficulty, but try to adapt the subject to the conventional names, as the object of this work is practical aid in treating diseases, no matter what they are called.

Fibræmia is where the fibrin is in excess in filaments, skeins, curled massive fibres like strings—thrombi and emboli. These are in a more exaggerated condition and form than in consumption or rheumatism, and are not necessarily associated with the crystalline matters or gravel. Sometimes the fibres look like a scalp that has been taken from the head of a woman with long tresses of hair.

G. THROMBOSIS

Is where masses of fibrin accrete and consolidate together, including or not the red corpuscles, white corpuscles, crystalline and pigmentary bodies, spores and mycelial filaments or vegetations, one or all.

H. EMBOLISM

Is where a thrombus has been caught or engaged in a blood-vessel and acts as a plug disturbing the circulation. When the embolus is made up of spores of mycoderma aceti or vinegar yeast and is caught in the lungs, it develops tubercle of the lungs, and so in other parts of the body. So senile gangrene of the extremities is caused by fibrinous clots plugging up an artery.

I. PRE-EMBOLIC STATE.

As thrombi precede emboli, so they can be detected in the blood before the embolism, simply by the morphology of the blood. In this way, sudden deaths from embolism, specially in the puerperal state, can be averted, and this aid alone renders the microscope an invaluable assistant to the physician who is devoted to his profession, and is sufficient to redeem it from the title of "accursed," as given it lately by a divine of this city.

J. ANÆMIA

Is where the serum is in excess and the red and white corpuscles are in diminution; fibrin also in excess.

K. PERNICIOUS ANÆMIA

Is where the red corpuscles are not formed or normally replaced. Here the blood glands are at fault, from improper alimentation. It is essentially a food disease.

L. MORPHOLOGY OF THE BLOOD IN SYPHILIS.

This morphology can be found associated with any of the preceding morphologies, but, when present by itself, it is recognized in the
Serum in two forms.
First. The spores of the crypta syphilitica.
Second. The mycelial filaments or full development of the same. The fructification is yet to be seen.

1. *The spores* are very minute, automobile, very lively, active, and saltatory. Carefully focussed a little off, they show a copper color. They dance about in the serum spaces and over the red corpuscles, where they elude search, unless one is a good and careful observer. They also crowd or are crowded into the white corpuscles, in which their color appears to greater distinctness, and which corpuscles are often distended to a great size.

2. *The mycelial filaments* of the crypta syphilitica are round, cylindrical, slightly tapering, mostly in small curved pieces broken off, with

one end larger than the other, or clavate at one end.

Color, when a little out of focus, copper. Sometimes they are long and wavy, sometimes branching. They are found in best condition in the walls of chancres.

The great value of a diagnosis of the morphology of syphilitic blood lies in the almost instant detection of the disease without a word to the patient, and in telling at once when the disease is cured, for it is not cured unless the blood is free from the plant.

The use of this morphology would prevent the terrible lesions of tertiary syphilis, as the patient would not be allowed to run into this stage. It tells at once the real progress of the case under treatment, and shows how remedies act, or if they are good for anything. It amazes the writer to see how indifferent the profession are to the morphology of syphilitic blood. It is an "idle tale," just as ocean steam navigation, telephony, and railroading were. Ere I die, I hope to see the world enjoying the benefit of this use of the microscope, as it does the once "idle tales" named.

M. MORPHOLOGY OF THE BLOOD IN ECZEMA.

Here the spores are black and still, not automobile, but passive. Parent vegetation not made out. This morphology may be found associated with any of the others. No case of eczema is cured unless these spores are eliminated.

N. MORPHOLOGY OF THE BLOOD IN SCROFULA.

This is either syphilitic or tuberculous, or both. See the morphologies of consumption and syphilis.

O. MORPHOLOGY OF THE BLOOD IN MALARIA.

Here the diagnosis rests on the forms found in the serum. There are:

1. *The spores* of the gemiasma plants or other plants found in malarious districts, which rise in the air from the soil, and are inhaled into the air passages where the blood comes within one-three-thousandth ($\frac{1}{3000}$) of an inch of the atmosphere. They there gain admission to the blood.

2. *The sporangias* of the mature gemiasmas.

These are pale or white in color, and generally contain less spores than normal, as would be expected in algæ growing in an unnatural habitat, as the inside of the human body.

Remarks.—1. Are most common. 2. Are rare, but in doubtful cases, if the skin morphology of the axillæ is studied, the full-grown aerial form of the gemiasmas may be found there for corroborative diagnosis. The malaria blood morphology may exist in a latent condition in persons apparently healthy, needing a torpid liver or a cold to make their energy actual, just as in the case of the loaded gun alluded to above.

There are several kinds of cryptogamic vegetations that cause malaria. Some of these are innocent vegetations in their natural habitat, but when animalized by coming in contact with animal matter in decay, and living on it, then they are endowed with a power to attack and live on the human habitat, and become the predisposing cause of malaria—so termed probably because these causative vegetations invade through the air—when taken into the digestive organs, as they must be in quantities, they seem to be destroyed by the juices of the alimentary canal. See " Malaria," McNaughton prize essay, 1882, by J. H. Salisbury, M.D., LL.D. New York: W. A. Kellogg, 1885.

P. HEREDITARY TAINTS

Are conventionally supposed to come through the blood, but the evidence of blood morphologies does not bear out this idea in a general way. Consumption comes by feeding on food that undergoes alcoholic and vinegary fermentation in the digestive organs.

The spores of crypta syphilitica and eczema may be transmitted from the mother or father to the offspring, but they are now about the only ones that can be traced.

Q. CANCER

Is more a disease of nutrition—tissue developed under mob law—and goes in families, because families feed on the same food at the same table. The researches of Dr. Domingos Freire, of Rio Janeiro, and others point out a microbe. This is an advance in our knowledge, for hitherto we have been able to detect no vegetation in cancerous blood before the general system has been broken down in the last stages, and here it seems more a result than a cause. But we are grateful for any more light, and accord Dr. Freire all credit and honor for his work.

R. MORPHOLOGY OF THE BLOOD OF VARIOLA AND VACCINIA.

Ios variolosa vacciola spores and filaments in variola.

Ios vacciola spores and filaments in vaccinia.

S. MORPHOLOGY OF THE BLOOD IN TYPHOID FEVER.

Biolysis typhoides spores and filaments.
The spores grow with great profusion in the

white blood-corpuscles, leaving them as empty sacs sometimes floating in the blood stream. The spores also grow in profusion in all the epithelia of the body. Patient not cured before the plant is removed.

T. MORPHOLOGY OF THE BLOOD IN SCARLET FEVER AND DIPHTHERIA.

Scarlet fever.—Mucor malignans spores, or a species very near kin to it.

Diphtheria.—Mucor malignans.

The aerial form may be cultivated from the throat membranes, but it is very dangerous work. The writer found that a three and a half years' sojourn of the diphtheritic membrane (from the uvula of his daughter Mary who died in spite of all that was done) in strongest carbolic acid was not enough to destroy the life of the vegetation. He confesses he was frightened, and abandoned the study of this particular spore.

U. MORPHOLOGY OF BLOOD IN FATTY DEGENERATION.

The white corpuscles contain globules of fat more or less abundant. The serum in advanced cases, or cases tending that way, contains fat globules more or less large and numerous.

The red corpuscles are apt to have not full

color, strength of outline, and be adhesive, pale, sticky.

Remarks.—The fibre of an outlying muscle may be brought out by a minute spear thrust in and tested for fat in the fibrillæ (S.), as a confirmation of the diagnosis. Very important in the treatment of softening of the brain, apoplexy, Bright's disease, etc.

V. THE MORPHOLOGY OF THE BLOOD IN FIBROUS CONSUMPTION.

Here the mycoderma aceti or vinegar yeast does not get into the blood, and change it, as in tubercular consumption, since the pylorus keeps the vinegar yeast in the stomach. There is breaking down of living tissue to a less extent. This tissue has been thickened, hardened, and made stony from deposit of gravel. The diagnosis is not so easy as that of tubercular consumption.

W. CHOLESTERÆMIA.

Red blood-discs soft, yielding, plastic, often sticky, holding feebly the coloring matter which escapes and obscures the field.

Serum contains cholesterin.

Diagnosis.—Blood standing a few hours on the slide; crystals of cholesterin appear on the edges of the slide.

This shows a tendency to amyloid disease in the spleen, lacteal and lymphatic glands, liver, kidneys, heart and large blood-vessels, and *amyloid matters are found in the blood stream.*

X. MORPHOLOGY OF THE BLOOD IN CARBUNCLES.

Crypta carbunculata spores and filaments which are found in abundance also in the sloughs of the carbuncle.

Y. THE MORPHOLOGY OF THE BLOOD IN YELLOW FEVER.

Cryptococcus xanthogenicus (Freire). See his monumental work.

Z. LEUCOCYTHÆMIA

Is where the white corpuscles are in large excess and the red corpuscles in diminution; serum in excess.

II.

MORPHOLOGY OF SPUTUM.

MODE OF STUDY.

One and one-fifth inch objective; one inch ocular.

Polarized light needed sometimes to distinguish the fibres of lung tissues from other organic fibres.

At least three specimens should be collected and studied at each examination. Sputum may be sent from patients prepared as follows: dry, away from sun or stove, a mass of morning sputum about one inch in diameter on white writing paper. The specimen will keep indefinitely and may be mailed anywhere. When ready for examination, soak specimen with a little water. The objectives made by the late Mr. Tolles and by his successor, Mr. John Green, will focus through a slide. It is therefore much easier to place some of the moistened sputum on a slide and then cover with another slide; this is done quicker than when one has to use thin covers. It is a pity that other American objective makers cannot follow the example of the illustrious Tolles, and make one-fifth inch objectives that will focus an eighth of an inch from the object, and not a sixteenth or thirty-second, as the common rule is.

Sputum needs morphological study as much as urine or blood.

As the morphology may include that of the air, of course this is incomplete.

Aerial forms of yeasts.
Albuminoid matters.
Alcoholic and lactic acid yeasts.
Algæ, names unknown.
Amorphous organic and inorganic matters, including dust and dirt inhaled from the atmosphere.
Amyloid bodies.
Anabaina irregularis.
Any of the microscopic fauna and flora found in drinking waters.
Asthmatos ciliaris.
Bacilli.
Bacteria, so-called.
Blood-corpuscles, white and red.
Butter.
Calculi made up of:
 Cholesterin.
 Cystin.
 Oxalate of lime.
 Phosphate of lime.
 Triple phosphates.
 Uric acid.

These may all come under the appellation of "gravel of the lungs."
Carbon, from smoke inhaled.
Carbonized tissue from lungs.
Cells and fibres of lung tissue.
Cholesterin.
Clots of blood.
Colloid.
Connective animal tissues.
Contents of giant cells escaped outside of walls.
Cotton fibre.
Cream of tartar crystals.
Crystals with two or more terminals.
Cystin.
Dust and dirt.
Elastic lung fibres.
Elements of animal food eaten, cooked and uncooked.
Elements of vegetable food eaten, cooked and uncooked.
Epithelia, ciliate, non-ciliate, pavement, columnar.
Fat.
Feathers.
Foreign substances inhaled.
Fucidium.
Fusiform crystals.
Gemiasma alba.
Gemiasma rubra.
Gemiasma verdans.

Granular tubercular masses.

Granular tuberculous matter, so-called, sometimes fetid in odor.

Gravel, crystalline.

Gravel, granular.

Gravel, massive.

Hairs of plants and animals.

Inelastic lung fibres.

Ipecac dust.

Lactic acid alcoholic yeast.

Lactic acid mother of vinegar.

Lactic acid vinegar yeast.

Leptothrix buccalis spores and filaments.

Leptothrix buccalis heavily loading and enormously distending lingual papillæ with spores and filaments.

Leptothrix buccalis in epithelia.

Linen fibre.

Lumina of blood-vessels.

Micrococcus spores.

Microsporon furfur.

Mucor malignans (diphtheria).

Mucous cells swarming with the moving spores, probably of the leptothrix buccalis; not found in the mouths of healthy infants.

Mucous corpuscles.

Mucous corpuscles, caudate and deformed.

Mucous corpuscles distended with albuminoids.

Mucous corpuscles distended with crystalline and other bodies.

Mucous corpuscles distended with cystin or giant cells.

Mucous corpuscles distended with leptothrix.

Mucous corpuscles distended with melanotic matters.

Mucous corpuscles distended with oxalate of lime.

Mucous corpuscles distended with triple phosphates.

Mucous corpuscles distended with uric acid and urates.

Mucous corpuscles, normal.

Mucous filaments and fibres.

Mucus; normal and ropy and viscid (colloid).

Muscular fibres of food.

Mycelial filaments of acetic acid vinegar, and lactic acid vinegar yeasts.

Mycelial filament of fully developed yeasts and other fungi.

Mycoderma aceti, spores and filaments.

Other crystals whose names have not been made out.

Oxalate of lime.

Papillæ of tongue, usually infiltrated with spores of leptothrix.

Partially carbonized vegetable tissues from smoke.

Phosphate of lime.

Pigment matters.

Pitted ducts, etc.

Portions of feathers of animals and insects.

Potato starch.
Pus-corpuscles.
Sarcina.
Silk fibre.
Skeins of mycelial filaments.
Special pollens.
Spirilina splendens (asthma), Salisbury, 1865.
Spirillum.
Spores of artemisia absinthium.
Starch, corn.
Starch, potato.
Starch, wheat.
Swarms of spores.
The whole lumen of a vein just before ending in the capillary.
Tough, ropy mucus.
Triple phosphates.
Tubercles.
Uric acid and urates.
Uric acid crystals.
Vegetable tissues.
Vegetations found in croupal membranes (Cutter, 1879).
Vibriones.
Vinegar yeast.
Vinegar yeast and lactic acid vinegar yeast.
Wheat starch.
Woody fibres.
Yeast plants.
Yeast sporangia, alcoholic and lactic acid.

III.

MORPHOLOGY OF FECES.

SHOWS THE CONDITION OF DIGESTION, GOOD, BAD, OR INDIFFERENT, AND SOME PATHOLOGICAL STATES.

MODE OF STUDY.

Prepare specimens for mailing, in the same way as sputum. A good microscope, one inch, one-fifth inch objectives, one inch ocular, polarized light.

Moisten specimen, and place on slide, and (if the physician has a fifth-inch objective that will focus through a common slide) cover specimen with a piece of slide. This is quicker, easier, cleaner, and more effective than with thin covers.

Acetic acid yeasts.
Alcohol.
Another species of sarcina.
Bacteria.
Beard of wheat.
Beef-red pieces of thickened mucus.
Black pigment from glands of Lieberkuhn and Brunner.

Blood.
Butyric acid yeasts.
Carbonate of lime.
Casts of intestinal glands.
Cholesterin.
Colloid.
Colloid matters, resembling ovarian, thyroid, and mammary tumors and those of testes.
Cotton fibre.
Cream-colored pus.
Crystals of phosphates, cystin, urates, oxalates, etc., colored with melanotic matters.
Crystals of sugar.
Crystals of triple phosphates.
Crystals of cystin.
Crystals, urates, uric acid, etc.
Different vegetable fibres.
Eggs of ascarides.
Eggs of different worms.
Eggs of tænia.
Eggs of trichocephalus dispar.
Epithelium.
Fat with acicular crystals.
Fat globules.
Gelatinous mucus.
Gluten.
Granular, amorphous, homogeneous matter, normal feces, with triple phosphates.
Healthy feces are homogeneous, formless, like a solid extract.
Homogeneous fecal matter.

Lactic acid yeasts.
Linen fibre.
Microcystis and plants allied to them, unnamed.
Mucous corpuscles.
Muscular fibre.
Mycoderma aceti.
Oil globules.
Oils.
Oxalate of lime.
Partially cooked and burnt muscular fibres.
Penicillium.
Pigmentine, black, etc.
Remains of animal tissues:
Connective tissue.
Striated fibres: striæ non-, partially or wholly effaced by digestion, etc., etc.
Remains of vegetable tissues:
Apples:—Clear, almost transparent sacs of thin cellulose.
Baked beans:—Sacs of thick cellulose containing starch cells; when un- or partially cooked, they are globular, pyriform, elongated, compressed, apparently triangular, sometimes reminding of difflugia cratera, sometimes of pelomyxæ, and so on; the transparent envelope of cellulose looks like the clear margin of gemiasma verdans, rubra, and plumba found in malaria. The thickness of this coat is about one-seventeenth of the diameter of the sac.
The starch cells polarize light or not as they

are uncooked or cooked. The cellulose envelope of the entire bean is made up of layers of crystal-like shapes, which are set in rows, their internal and external faces appearing very much like the tops of the Giant's Causeway crystals of traprock. These crystal-like elements of cellulose, when un- or partially cooked, are but slightly hourglass-shaped, but, when thoroughly cooked, appear like double-headed tacks.

Epithelial cells and areolar tissue of beans may also be present.

Bananas:—Clustered masses of starch grains.

Cranberries:—Pigment cells of skins.

Greens:—Spiral ducts in bundles, etc.

Potatoes:—Cork cells, starch cells, areolar tissue. Gubernaculum tissues that lead from the eyes to the centre. The starch bundles or the starch in homogeneous masses, the pitted ducts, the vascular bundles, etc.

Wheat: beard, outer coats, gluten cells, areolar tissue, etc., etc.

This is only a very partial list of vegetable tissues. I have only indicated a few elements in order to show how to go at the study, for my own work has led me to distinguish many more forms.

Saccharomyces cerevisiæ.

Sarcina ventriculi.

Seroline.

Several species of minute algæ.

Shreds of coagulated mucus.
Sirocoleum.
Strings of thin folded laminæ of coagulated mucus.
Strips of tissues, scourings.
Sugar.
Sulphuretted hydrogen vegetations.
Tarry condition from bile which should have been carried out by urinary organs and sweat glands (Salisbury).
Tegument of wheat, cigar coat.
Tough ropy mucus (colloid).
Triple phosphates.
Tubercles.
Urates.
Vegetations of putrefactive decomposition.
Vinegar.
White coagulated mucus, like folded tissue paper.
White connective fibrous tissues.
Yeast plants.
Yeasts:
    Acetic acid.
    Alcoholic.
    Butyric acid.
    Lactic acid.

IV.

MORPHOLOGY OF THE SKIN.

MODE OF STUDY.

Simply moisten the skin with distilled water and rub in with a clean knife blade. Then scrape off and place under microscope; use one-fifth inch or one-tenth inch objectives or higher as needed, having water enough to make a thin clear field; in studying dirt and some of the grosser forms, use lower powers.

Acarus autumnalis.
Acarus folliculorum, steatozoon folliculorum.
Acarus scabeii.
Acne.
Adenoid.
Ague plants. Among these gemiasma alba, gemiasma plumba, gemiasma rubra.
Anabaina subtularia.
Animal hairs.
Anthrax vegetations.
Asthmatos ciliaris.
Bacteria.
Blood, free and dried.
Blue, purple, black pigments.
Boils, vegetation of.

Bots.
Cancer.
Carbonate of lime.
Carbuncle, anthrax.
Carpet fibres.
Chloasma.
Cholesterin.
Cimex lectularius.
Crypta syphilitica (Salisbury) spores and filaments.
Cystin.
Dermatophyton.
Dirt.
Drugs, ipecac, etc.
Eczema spores.
Eggs and larvæ of insects.
Eggs of tape-worm.
Epithelia, normal.
Epithelia, lactic acid yeast in.
Epithelia, vinegar yeast in.
Epithelia with biolysis typhoides.
Epithelioma.
Erysipelas vegetations.
Fat.
Fatty degeneration.
Fatty infiltration of muscles.
Favus, tinea favosa.
Feathers.
Fibres of textile products, cotton, linen, wool.
Fibroid tissues.
Filaria medinensis.

Floor fibres.

Flour and flour vegetation, as on baker's wrists.

Fungi.

Fungoid spores and mycelia of unnamed plants.

Germs in epithelial and mucous tissues, glands and follicles of eye and other organs.

Gravel, foreign and native.

Hairs and vegetations.

Jiggers.

Jute.

Keloid.

Kerion.

Lard.

Leather.

Leprosy spores.

Lice.

Malignant pustule vegetations.

Measles vegetations.

Mentagrophyton.

Microsporon Audouini.

Microsporon furfur.

Mosquitoes, parts of.

Mucor malignans of scarlet fever and diphtheria.

Mucus.

Mycetoma, Chionyphe Carteri.

Mycoderma aceti spores and filaments.

Nails, vegetations and dirt under.

Oils.

Onychomycosis, onychia parasitica.
Oxide of lime.
Paint lead salts.
Pediculus capitis.
Pediculus corporis vel vestimenti.
Pediculus pubis.
Phosphates of lime.
Pigment matters.
Plant hairs.
Poisonous plant products.
Pollen of plants.
Porrigo scutulata or tinea tonsurans.
Protococcus monetarius under ends of finger nails.
Pulex or sarcopsylla penetrans, Chigoe.
Pus.
Pus decomposing into fat.
Saccharomyces cerevisiæ.
Salt, chloride of sodium.
Scald head.
Scarlet fever vegetations, mucor malignans.
Scars of pregnancy and fat distention.
Seborrhœa.
Secretions of hair and sweat glands.
Serum.
Silica.
Silk.
Small-pox vegetations.
Soap.
Spermatozoa.
Stains of silver, etc.

Starch grains of all kinds.
Steatozoon folliculorum.
Stellin.
Stellurin.
Sulphate of lime.
Sweat.
Syphilodermata.

The objects found in the *morphology of the air* are to be expected in the morphology of the skin.

Tinea circinata, trichophyton tonsurans.
Tinea decalvans, microsporon Audouini.
Tinea favosa, achorion Schonleinii.
Tinea kerion.
Tinea sycosis, microsporon mentagraphytes.
Tinea tarsi, tricophyton.
Tinea tonsurans, tricophyton.
Tinea versicolor, microsporon furfur.
Trichosis caninus (Salisbury).
Trichosis felinus (Salisbury).
Triple phosphates.
Uric acid.
Vaccinia vegetations.
Variola vegetations.
Vegetations from water used in washing.
Vegetations of animal poisons.
Vibriones.
Woody fibre.
Yeasts growing in epithelia of skin.
Zinc, oxide of.

V.

MORPHOLOGY OF THE URINE.

It is good to use an inch objective as well as a fifth (1-5) inch objective in studying the morphology of the urine. The one-inch objective at once brings out the casts of kidney tubes, prostate gland ducts, spermatic ducts, besides the colloid matters that otherwise elude search and are, in my opinion, very important clinically.

Urinoscopy is more valuable than the pulse in telling the status of the liver, stomach, kidneys, urinary organs, and general systemic condition. It should be used daily. The urine voided on rising in the morning is the best to examine. *The chemical examination of the urine should go side by side with the morphology; neither supersedes the other.*

The aim should be to make each patient's urine come up to the standard of the urine of a healthy infant, nursing a healthy mother's breast. *This urine is clear, odorless, and free from deposit.* The cures in the so-called Salisbury plans include an aiming at a conformity to this standard. It is a mistake for each physician not to make his own examinations of urine almost daily. The urine is very sensitive to bad feeding and overdoing in any way, and shows them almost as plainly as if it said "bad feeding and overdoing" in so many words. Lastly, many physicians will not examine urine chemically or microscopically, as such examinations appear to be too difficult, though these men may be masters in the practice of medicine; there is nothing in the present knowledge of the urine that any one of moderate ability may not and should not master, for, to repeat, the urine is a source of valuable clinical information.

Accidental foreign bodies.

Acicular crystals, same as found in ague soils.

Ague plants, mostly in embryonic forms, sometimes mature.

Amorphous urates.

Amyloid matter, common.

Anabaina irregularis.

Arachnida.

Asthmatos ciliaris (rare).

Bacilli.

Bacteria.

Blood, red corpuscles.

Blood, white corpuscles.

Calculi of urates and phosphates from pelves of kidneys or not.

Cancer cells must not be mistaken for giant cells with prolongations sometimes ten times their length, and sometimes connecting with gubernacula two or more giant mucous cells.

Carbonate of lime.

Casts of spermatic ducts, clear or with amyloid, phosphate of lime, triple phosphates, etc.

Catarrhal discharges from spermatic ducts or the prostatic glands:
>   (1) Protoplasmic.
>   (2) In skeins.
>   (3) In Indian clubs.

These occur together at times; a supposed cause of neurasthenia in men (Cutter).

Chyme.

Colloid matter.

Cotton, wool, bast, linen fibres, indeed any form from the morphology of the air may get in accidentally.

Crypta syphilitica spores.

Cryptococcus xanthogenicus.

Crystals with radiations formed within cells with amœboid projections.

Cystin.

Dirt.

Dust.

Eggs of ascarides.

Eggs of trichocephalus dispar.

Epithelia invaded by vegetations of scarlet fever, diphtheria, typhoid fever, etc., etc.

Epithelia, pavement and columnar, from bladder and vagina.

Fat in globules.

Fatty casts of kidney tubes.
Fatty epithelia from kidneys.
Fragments of animal and vegetable tissues.
Giant cells distorted and connected together by gubernacula—parent mucous cells, probably simulating cancer cells.
Gemiasma rubra.
Gravel.
Hyaline casts of kidney tubes.
In perfect health, free from deposit or odor, like healthy nursing infant's urine.
Lactic acid yeasts, spores and filaments.
Mucous cells.
Mucous fibres and casts from kidneys.
Mucous filaments.
Mycelial filaments of mycoderma aceti—sometimes mother of vinegar.
Other algæ.
Oxalate of lime, granular and in dumb-bell.
Penicillium.
Phosphates.
Phosphate of lime.
Pigment matters.
Pus cells.
Putrefactive yeasts in spores and mycelial filaments. When these are voided from the bladder, in spores single or aggregated, filaments single or in skeins, I regard it as a diseased condition, to be treated as such. Have known epilepsy to be caused by them, and cured by their removal (Cutter).

Radiating plants, same as found in ague soils.
Saccharomyces cerevisiæ, or alcohol yeasts.
Spermatozoa, normal.
    With two heads.
    With three heads.
    With two tails.
    With three tails, etc.
    With two heads and two tails.
    With three heads and three tails, etc.
Sphærotheca spores and filaments.
Starch grains.
Triple phosphates.
Urates of soda and ammonium.
Uric acid.
Vegetations of gonorrhœa.
Vibriones.
Waxy casts of kidney tubes.
Yeasts.
Zooglœa forms.

VI.

THE MORPHOLOGY OF THE VOMITUS.

Any object of the morphology of foods.
Bile.
Blood.
Butyric acid fermentative vegetations.
Cancerous matters.
Chyme.
Coagulated food.
Colloid.
Cryptococcus xanthogenicus.
Epithelia.
Food partly digested.
Lactic acid yeasts.
Morphology of feces, rare.
Mucous corpuscles.
Mucus.
Mycoderma aceti.
Saccharomyces cerevisiæ.
Sarcina ventriculi.
Slime.
Sometimes yeast plants form a coating on œsophagus, discharged as a membrane.

VII.

MORPHOLOGY OF FOODS.

A. WATERS OF LAKES, PONDS, AND WATER SHEDS; HYDRANT WATERS. *

Morphology of animals, plants and other substances found in hydrant waters and pond waters, such as are used for drinking purposes.

The list is very incomplete, as more than half of the objects found have no names (Professor Paulus F. Reinsch, Erlangen, Ger.).

Over thirty hydrant waters of cities and towns were studied. Among these were those of Albany, Brooklyn and New York, New York; Arlington, Boston, Cambridge, Charlestown, Haverhill, Charles River, Jamaica Pond Boston, Lynn, Malden, Salem, Springfield, Winchester, Woburn, Worcester, Wellesley Hotel, Massachusetts; Philadelphia, Penn.; Hartford and New Haven, Connecticut; Chicago, Illinois; Washington, D. C.; Dover, New Hampshire; Baltimore, Maryland; Cleveland, Ohio; Richmond, Va. Besides ponds in Amherst, Falmouth, Natick, Holbrook, Wake-

---

* See page 81 for mode of examination.

field, West Falmouth, Wellesley, Massachusetts; East Greenwich, Rhode Island, and North Turner, Maine.

LIST.

A beautiful entomostraca, like the branchippus stagnalis. (Croton.)

A delicate animal which looks like a snail, and yet without the terminal of the spiral. It is beautifully transparent, so that the motion of the heart is more apparent than in the following. (Croton.)

A magnificent animal composed of a hyaline sac open at one end. Transparent. Mouth provided with cilia, which are inverted completely within the body at will. The viscera are held together by gubernacula just outlined enough to be visible. These contract, and keep the viscera moving to and from the mouth; specimen name unknown to me; have found it only in the Croton.

Abundant mycelial fungus filaments.
Acineta tuberosa.
Acropherus.
Actinosphericum Eichornii.
Actinodiscus.
Actinophrys sol.
Alcyonella.
Alonia.
Amblyophis.
Amœba proteus.
Amœba radiosa.
Amœba verrucosa.
Amphiprora alata.
Anabaina circinalis.
Anabaina subtularia.
Anguillula fluviatilis.
Ankistrodesmus falcatus.
Ankistrodesmus unicornis.
Anurea longispinis.
Anurea monostylus with ovary one-half the diameter of its own body.
Anurea stipitata.
Aptogonum.

Arachnida.
Arcella mitrata.
Arcella vulgaris.
Argulus.
Arthrodesmus convergens.
Arthrodesmus divergens.
Arthronema.
Astrionella formosa.
Bacteria.
Bosmina.
Botryococcus.
Branchippus stagnalis.
Bursaria.
Campanularia.
Campascus carnutus.
Carapace of a monostyled rotifer, occupied by a parasitic mother cell with protoplasmic contents in very active motion. (Croton.)
Castor.
Centropyxis.
Centropyxis acuelata.
Chetochilis.
Chilomonas.
Chlorococcus.
Chlorogonium.
Chroococcus chalybeus.
Chydorus.
Chytridium.
Cladophora.
Clathrocystis æruginosa.
Closterium didymotocum.
Closterium lunula.
Closterium moniliferum.
Cochliopodium bilimbosum (Harriman).
Cœlastrum sphericum.
Confervoideæ.
Cosmarium binoculatum.
Cosmarium crenatum.
Cosmarium tetrophthalmum.
Cosmarium margaritiferum.
Cristatella mucedo.
Cyclops quadricauda.
Cyclops quadricornis.
Cyphroderia ampulla.
Cypris tristriata.
Daphnia pulex.
Desmidium.
Desmidiaceæ.
Diaptomas castor.
Diaptomas castor with saprolegnia attached.
Diaptomas, new species.
Diatoma vulgaris.
Didymocladon.
Difflugia cratera.
Difflugia globosa.
Difflugia lobestoma (Harriman).
Difflugia pyriformis.
Dinobryina sertularia.
Dinocharis pocillum.
Dirt.

Docidium.
Eggs of bryozoa.
Eggs of entomostraca.
Eggs of plumatella.
Eggs of polyp.
Empty shell of arcella.
Enchylis pupa.
Enteromorpha clathrata.
Eosphora aurita.
Epithelia, animal.
Epithelia, vegetable.
Eradne Nordmanni.
Euastrum.
Euglenia viridis.
Euglypha.
Eurycercus lamellatus.
Exuvia of some insects.
Feather barbs.
Fish scales.
Floscularia.
Fragillaria.
Fungi.
Fungus, red water.
Gammarus pulex.
Gemiasma verdans.
Globar rotifer.
Gomphospheria.
Gonium.
Grammatophora.
Gregarina sænuridis.
Gromia.
Hairs of plants.
Hairs of various animals.
Heleopera picta.

Holophrya brunnea.
Humus.
Hyalosphenia tincta.
Hyalosphenia formosa.
Hyalotheca.
Hyamodiscus rubicundus.
Hydra vulgaris.
Hydra viridis.
Infusoria.
Insect scales.
Lacinularia.
Lacinularia socialis.
Leaves and parts of leaves.
Leptothrix.
Leucophrys patula.
Licomophora.
Lyngbya.
Masses of sponge parenchyma decomposing.
Melosira.
Meresmopedia.
Micrasterias digitata.
Micrasterias denticulata.
Micrasterias rotata.
Microcoleus.
Milnesium tardigradum.
Monactinus octenarius.
Monactinus duodenarius.
Monads.
Mycoderma aceti.
Navicula amphirynchus.
Navicula cuspidata.
Nebalia bipes.
Nitzschia.

Nostoc communis.
Notodelphys
Oedogonium.
Oscillatoriaceæ.
Ovaries of entomostraca.
Palmellæ.
Pamphagus mutabilis.
Pandorina morum.
Paramecium aurelium.
Pediastrum boryanum.
Pediastrium incisum.
Pediastrium perforatum.
Pediastrum pertusum.
Pediastrum quadratum.
Pediastrum tetras.
Pelomyxa.
Penium.
Peridinium candelabrum.
Peridinium cinctum.
Phacus.
Plagiophrys.
Plagiotoma lumbrici.
Pleurosigma angulatum.
Plumatella.
Pollen of pine.
Polyartha platyptera.
Polycoccus.
Polyhedra tretætica.
Polyhedra triangularis.
Polyhedrium.
Polyphema.
Polyphemus pediculus.
Protococcus.
Protococcus viridis.

Radiolaria.
Radiophrys alba.
Raphidium duplex.
Rotifer ascus.
Rotifer vulgaris.
Saccharomyces cerevisiæ.
Saprolegnia.
Sarcina.
Scales of butterfly.
Scaridium longicaudum.
Scenedesmus acutus.
Scenedesmus obliquus.
Scenedesmus obtusum.
Scenedesmus quadricauda.
Setigera.
Sheath of tubularia.
Silica.
Sphærotheca spores.
Spicules of sponge.
Spirogyra.
Sponges.
Starch.
Staurastrum dejectum.
Staurastrum furcigerum.
Staurastrum gracile.
Staurastrum margaritaceum.
Staurogenia quadrata.
Stephanocerus.
Stephanodiscus niagaræ.
Spiral tissue, etc.
Spirotænia.
Stentor.
Surirella bifrons.

Surirella gemma.
Synchœta.
Synhedra.
Synhedra splendens and many other diatoms too numerous to name.
Tabellaria.
Tetmemorus granulatus.
Tetraspore.
Trachelomonas.
Triceratium favus.
Trichodiscus.
Tryblionella scutellaria.
Ulothrix mucosa.

Urococcus.
Urostyla.
Uvella.
Vegetable fibres.
Volvox cœnochilus.
Volvox globator.
Volvox, new species.
Vorticel.
Wheat starch grains, etc.
Worm fluke.
Worm, two tailed.
Xanthidium.
Yeast.

APPENDIX.

The bad taste in Cochituate, cause of, discovered in 1879, and in Croton water, discovered in 1881, to be due to the presence of spongilla fluviatilis and lacustris and the pelomyxas.

The following facts are adduced in support of this belief:

1. Spicules of sponge were very abundant in the Croton during the time of bad taste. These spicules are most elegant forms of silica that will not polarize light. They are of various shapes. The most common one is that of a boomerang shape, exquisitely pointed at both ends, and polished like steel. Another common form in the Croton is shaped like an old-fashioned two-tined fork such as is used in a pork barrel. Some are like the little stand used on the dining-room tables to keep the blade of the carving

knife off the cloth, etc., etc. Now, this sponge itself is made up of a jelly-like substance, or sarcode protoplasm. When the animal dies, the sponge jelly or protoplasm is dissolved in the water, and goes through all filtering apparatus. For example, in Woburn, Mass., the hydrant water is taken from a gallery by the side of Horn Pond. Though this water is clear as crystal, if the nostrils be placed over a goblet of it, only a few sniffs are necessary to perceive the peculiar earthy smell, though the sense of taste detects nothing wrong. (Parenthetically, a firm brought suit against this town for loss of water power by the use of this spring for drinking purposes. My testimony, that the two waters were identical by morphological examination, helped the case for those suing for damages.)

Now an abundance of spicules shows an abundance of sponges. When Professor Reinsch and myself were studying the Cochituate water, it was a great problem to find the sponges from whence these spicules came. It always seemed to me that the minute spongilli, as found on the rocks of the bottom of ponds, did not adequately explain the presence of the sponge spicules, so I kept watching for them, and was rewarded in 1879.

2. I found in Charles River (Mass.) a fresh-water sponge that was as thick as my little finger, and between three and four feet long in linear measurement. Also a clustered mass of sponges in the same river large enough to fill a two-bushel basket.

Officials connected with the Boston Water Works have informed me that they have seen like collections of sponges in the sources of the Cochituate water. From finding the spicules so abundant in the Croton, I inferred that there is an abundance of the same sponges in the sources of the Croton water supply.

3. A portion of the Charles River sponge kept over night in a tumbler, in my room at the Wellesley Hotel, stank, in an exaggerated measure, to be sure, as the filter stank after filtering the Croton water, winter of 1880–81.

4. About January, 1881, Dr. Harriman, of Boston, my associate, and myself found portions of dead and decaying sponges in the Cochituate, they not having been dissolved. Some of the spicules were actually sticking out of the mass. The Cochituate had as bad a taste, and worse, than the Croton at that time.

5. As said before, the great mass of the dead sponges are soluble in water, and go through all filters.

It seems to me reasonable to partly attribute the taste and smell to which allusion has been made, to the presence of sponges. They die and dissolve in the water, and were it not for the tremendous draft on the supply, would, no doubt, be all disposed of by the plants and scavengers living in the water. I am aware there are some difficulties in the way of this explanation, from the fact that we find sponge spicules at all seasons of the year, and why, then, should not the taste be bad all the time? In reply to this, I refer to the abundance of the spicules being greater at the time of the worst taste. I would not be understood as claiming that the dead sponges are the sole cause of this taste, as there are a great many rhizopods (root-footed animals) in the water that die also. They are protoplasmic, like the sponges. They die, but leave more solid remains than the sponges. Dr. Harriman and myself have noticed especially the pelomyxas (pelos, mud, and mukos, mucus) animals made up of a jelly-like protoplasm, that are very greedy. They are figured in Dr. Leidy's magnificent work on the rhizopods, issued by the U. S. government. We have found them very abundant in the Cochituate and the Croton when this bad taste was most palpable. Now as to the question whether the drinking of water impregnated with dead sponges is healthy. I am sure no one would have wished to drink the water I had in my room at Wellesley, fetid with dead sponge; but as to the Croton, the chemists decide, I understand, that the drinking of dead sponges and pelomyxas is not and cannot be a cause of disease. Now the dicta of the chemist must be respected, as we have said, and always shall say; but when it comes to a question so subtle as the causes of disease, as a physician I should hesitate before I pronounced definitely on the question, for the reason that there is such a great difference in people as to food. Some people will eat food with impunity that in other cases acts as a poison to others.

Again, the question of the causes of disease is by no means settled, and it will be a long time before there is an agreement. For example, take consumption. I believe in the Salisbury plan, that it is a disease primarily of the blood, caused by the vinegar yeast. Though this view is supported by the synthesis of the disease in hundreds of healthy animals killed by feeding on yeast plants, and the disease verified by examinations after death, by micro-photography of the forms in the blood, and by the cure of a large number of persons, still very few of the profession have received this view, and have expressed no opinion about it. So that supposing, for example, the question should be raised, if the dead sponges in the Croton water could cause consumption by introducing the vegetations of decomposition into the human system, I think a chemist would shrink back from it into his laboratory, as it would be so difficult, in the present state of knowledge in the medical profession, to have the expression of a decided opinion.

How does the chemist know that dead sponges do not cause disease? Diseases do exist, but their ætiology is not found in the books of chemistry.

As a physician, I say that the question is still sub judice. To solve it will require the combined action of the zoologist, the botanist, the pathologist, and the practical physician. "But," you say, "we cannot wait for this; what shall we do until the question is decided?" If a reply is forced, I should say it would be a very sensible precaution to filter and boil the water when it tastes badly. The labors of Professor Reinsch have proved that cotton is king as a filter. This royal gift is common everywhere.

B. WATERS OF WELLS AND SPRINGS, UNCONNECTED WITH PONDS OR LAKES.

Spring Water from the Farm of Mr. George Plum, Mantua, Ohio.

Bacteria.
Diatoma vulgaris.
Epithelia from vegetables and animals.
Feather.
Linen fibre.
Mass of vegetable cells, probably of some berry.
Protococcus.
Silica or sand.
Small masses of dirt.

Sphærotheca, a fungus spore.
Starch.
Woody fibre.

Fitchburg Gas Company's Water, Specimen Furnished by Miss E. W. Beane, Teacher, July 16th, 1881.

Bacteria, few.
Cotton fibres.
Epithelial cells.
Leptothrix.
Linen fibres.
Mycelial filaments of a small water fungus.
Tabellaria.

This water has a high local reputation, and if the present morphological examination is verified by several more examinations, it must sustain a very high, if not the highest, reputation as a drinking water for the public. Here the work of filtering is done by the everlasting hills. It is an instance where the nearest approach to perfection in filtering is seen, provided the specimen sent is an average sample.

Water from Iron Tube Driven Well, West Falmouth, Mass. My own.

A few bacteria.
Oil globules.
Particles of dirt.

Pavement epithelia from human skin, probably came from the contact of a sewer's fingers who made the cotton filter.

Scales of oxide of iron.

Starch grains of wheat, that may have come from the new cotton-cloth filter used.

Epithelia made up most of the organic forms. The white cotton filter was stained red with the iron. Depth of well, fourteen feet. Soil, sandy. Location, within a hundred feet of the shore of Buzzard's Bay. Water saltish, but very cool and palatable. Supply, unfailing. It looks well to the eye.

W. A. Howland's Well, same place, Tubular.

Epithelia.

Large vegetable cell, transparent and surrounded by a flat ring.

Ditto, reminding of a cell of orange pulp.

Little dirt.

Mycelial filaments.

Oil globules.

Organic globule.

Starch.

This is nice water, and has agreeable effects on all the senses. It is down in the cellar of the house, and is about fourteen feet deep.

Water from Capt. Hoxie's Well; Has a Dead Animal Taste.

Bacteria, abundant.
Dirt, very abundant.
Epithelia in large collections.
Feathers.
Leather from new valve of pump.
Mass of decaying animal matter; dangerous water.
Monad, alive.
Mycelial filaments of yeast.
Organic globule, unknown.
Oxalate of lime crystal.
Silica.
Starch.

This was a common well, quite deep, and large enough for a man to get into. Comes through a lead pipe. Family sick and feeble.

The most striking result is the comparatively small presence in the springs and wells, namely, of organic forms of life as compared with the ordinary ditch, pool, or pond water. Still the fungi found may be more deleterious to health than all the forms in Croton, for example. This is what we are searching for. A member of the family using the well of Capt. Hoxie has had the pretubercular stage of consumption, as shown by physical micrographical explorations. Also his sputa, urine, and feces have been obstinately loaded with vegetation till lately. We are inclined to think this water has had something to do with it, and it will be prohibited. The sputa and kidney secretions kept for a day would be disgustingly fetid, while both would be loaded with vegetation, excretal. I never had so obstinate a case before. Neither diet, sulphur bathing, salicin, or quinine seemed to affect the abundance of the vegetation until after three months. Only after the inhalation of liquid ozone, of Parke, Davis & Co., of Detroit, did the vegetative solids disappear, but I have no doubt they would reappear if the use of this water is continued. I have never met with such an obstinate case (epileptic, etc.) under the use of hydrant drinking waters. Have had one case where the urine was loaded with vegetation as it left the body. This was a Croton-water drinker, but diet alone speedily removed the vegetation.

Water from the well of the late J. F. Davis, W. Falmouth, Mass.

Bacteria.
Cotton and wool.
Dirt, abundant.
Epithelia, in abundance.
Fungus, spores, ditto, sprouting.
Leptothrix.

Mycelial filaments very abundant on culture twenty-four hours.

Woody fibres.

It was said that the chemist's examination pronounced this well water perfectly pure. We are not prepared to say that they, the fungi, caused the sickness in question, but unhesitatingly advised the disusing of the water, for, as Dr. Harriman, my associate, said, "this abundant presence of fungi shows the presence of animal matter. At the same time the result shows the truth of the positions maintained here, that chemical exploration alone is insufficient for the examination of potable water."

C. ICE.[*]

MODES OF STUDY.

1. A clean bag, one inch by four inches, made of cotton cloth, was tied to the escape pipe of a refrigerator, zinc lined, shelf at top, that had been washed and cleansed with filtered water. The filtrate of from thirty to forty pounds of ice was collected by inverting the detached bag into a clean goblet, then sopping the inverted bag in the filtrate, and wringing the bag also.

2. A common silver ice pitcher, porcelain lined, was cleaned with filtered Croton water and

---
[*] See Scientific American of July 29th, 1882.

filled with broken ice, source unknown, clear, compact, solid, diaphanous, and pure looking. This was allowed to melt, and one quart of water resulted, and was filtered as before.

Power of microscope, one-fifth inch objective. Eye-piece, one inch and half-inch, 350 diameters.

Many of the following list come from the air; perhaps half. Some of the specimens of ice came from ice wagons; one from a provision store. This is, of course, a partial list.

Acanthodinium, with clusters of twelve spiral cells separated in all directions.
Actinophrys sol.
Alcohol yeast.
Amœba, alive.
Anurœa monostylus.
Ascus.
Astrionella formosa.
Bacillaria diatom.
Bacteria.
Bast fibres.
Botridium cells.
Broken down tegument and substance of leaves.
Bryozoa, egg of.
Carbon.
Chitin.
Chlorococcus.
Claw of water spider.
Claws of insects.
Closterium.
Closterium lunare, dead.
Closterium, young.
Coal.
Cœlastrum sphericum.
Collection of liber fibres.
Cotton fibre.
Corn starch.
Cryptomonas lenticularis.
Daphne claws.
Dark-red organic unknown body.
Decaying leaves.
Desmid, penium.
Diatoma, not named.
Diatoma vulgaris.
Diatomaceæ, other.
Difflugia.
Difflugia, dead, several varieties.
Difflugia globosa.
Difflugia, unusual.

Dinobryina sertularia.
Dirt, debris, etc.
Dust and excrementitious matters.
Egg of the fresh water polyzoa named below, unhatched.
Eggs of entomostraca.
Epidermis of wheat.
Epilobium montanum pollen.
Epithelia, animal and vegetable.
Epithelial scales, human.
Euglenia viridis.
Euglypha.
Euglypha cristata.
Exuvium.
Feather barb.
Fibre of wool colored blue.
Fish scales.
Foot stalks of vorticells, twenty-five in number.
Fungi and spores.
Fungus filament.
Gemiasma verdans.
Gluten cells, wheat.
Gromia.
Hairs of plants.
Hairs of various animals.
Humus.
Large double body, probably eggs, but possibly vegetable.
Large masses of decaying vegetable substances.
Large paramecia.
Leaves of moss.
Leptothrix.
Liber fibres.
Linen fibre imbedded in a mass of decaying vegetable substance.
Linen fibres.
Lyngbya.
Mass of carbon.
Melosira.
Membrum disjectum of a large entomostraca.
Monads.
Mycelial filaments, abundant.
Mycelial filaments, collection of.
Mycelial filaments of red water fungus.
Navicula.
Nebalia.
Nostoc.
One gonidia of coelastrum sphericum.
Oscillatoria.
Parenchyma of leaf.
Parenchyma of wheat.
Pavement epithelia, five specimens.
Pediastrum boryanum.
Pelomyxas, other.
Peridinium cinctum.

Peridinium spiniferum.
Piece of a red cranberry skin.
Pitted ducts.
Polyzoa.
Portion of a leaf with chlorophyll attached, color unchanged.
Portion of a red water fungus.
Potato starch.
Protococcus.
Protococcus, probably gemiasma.
Rotifer.
Scenedesmus obliquus.
Scenedesmus quadricauda.
Shell of a cyprus.
Silica.
Silk fibre.
Skeleton of leaves.
Sphærotheca fungus.
Spiral tissues of leaf.
Starch of corn, wheat, and potato.

Staurastrum.
Supposed egg of an entomostraca.
Tabellaria.
Tetraspore.
Trachelomonas.
Transverse woody fibre.
Vegetable epithelial collection.
Vegetable hair, long.
Vegetable hairs.
Vorticell, dead.
Vorticella, two joined together.
Wheat gluten cells.
Wheat starch.
Wood fibre of various kinds.
Wool.
Worm.
Yeast, alcohol, vinegar, and lactic acid.
Yeast, vegetating filaments.

Ice from Horn Pond, Woburn, Mass. This presented considerable lightish colored deposit, in which a few animal and vegetable forms were found, but was mainly made up of epithelia and amorphous dirt. The result was unexpected, as unfiltered Horn Pond water is rich in forms of life.

APPENDIX.

In this article of mine in the *Scientific American*, as before noted, there were illustrations to the number of eight. I give, as follows, some of the descriptive text of those illustrations:

Yeast.—This is the alcohol yeast of the yeast pot, torula cerevisiæ, the spores of which are everywhere present, ready to germinate if they have the opportunity. Its presence in ice is interesting.

Bacteria.—These are minute, self-moving protoplasmic bodies. Some regard them as ultimate forms of life; others that they are but the embryonal forms, seeds, or babies (as it were) of a vegetation, yet capable of immense reproduction by division, arranging themselves into masses, chains, etc., at will. In order to know what plants they belong to, culture is necessary. It is possible that those in the cut may be the spores or seeds of the yeast plants, but it cannot be said with certainty.

Pelomyxa.—This means "mud mucus." It is an animal classed with the rhizopod or root-footed protoplasmic animals. They are very greedy, and eat much mud or dirt. The color in this case is dark amber, and may be mistaken for decaying vegetable matter. The writer regards them with suspicion, as contributing, when dead and decaying, to cause the "cucumber" and fish-oil taste that sometimes occurs in hydrant drinking waters, notably the Cochituate.

Portions of Difflugia.—These are like the pelomyxæ, only they have the property of building over themselves a covering made of particles of sand, glued together so as to protect their structural protoplasmic bodies. Lately, the writer saw a difflugia cratera, whose shell had been broken on one side. The cilia that were usually seen at the natural opening were seen to be active at the artificial opening. The contour of the hole changed under view from circular to a narrower one, forming a segment of the first, showing an action of repair; suddenly there was a gush of protoplasmic jelly, and the animal was dead, dying in its efforts of reconstruction.

Mycelial filaments of a red fungus, found commonly in Horn Pond, Woburn, Mass. Also at Cambridge. Name not known to writer, nor Prof. Reinsch.

A curious dark-red tubular body, fragments of which I have often seen in hydrant drinking waters. Its fracture is glassy. It is an animal substance probably, and this is the best specimen I have seen.

Trachelomonas.—These are by Ehrenberg claimed as infusoria. They are very abundant in hydrant waters at all seasons of the year. The specimen here is dead, but the living individual moves its curious long flagelliform filament, by means of which it gracefully propels itself in any direction at will.

Astrionella Formosa.—A beautiful, very common diatom, that arranges itself into forms like the spokes of a wheel. Three spokes only are given here; usually twelve. This power of self-symmetrical arrangement is surprising and mysterious.

Bast or Linen Fibre.—This probably came from some table cloth, towel, or clothing.

An ascus or theca of a fungus, which is a part of a fructification of the fungus, and also found in lichens. It is strikingly well-developed.

Epithelia, probably animal.—These are suspicious organisms. See New York *Medical Record*, April 8th, 1882.

Egg of a bryozoa or polyzoa, found not unfrequently in the drinking waters of our cities and towns. It corresponds to the "winter egg" of entomostraca. It forms one of the four modes of reproduction, which Smith distinguishes: First. Eggs from spermatozoa. Second. From internal development (this very one). Third. External buds. Fourth. Brown bodies in empty eggs. This particular egg is seen to have an oval opening, whence the contents have been hatched or destroyed. It has been traced to a single polyp. Usually the animals live in a colony, and are met with in fresh water on stones, sticks, sides of flumes, and free. I have seen colonies of these bryozoa in masses as big as a bushel basket, hanging on and covering the perpendicular boards of a flume. In the present case, the egg is nearly as large as the animal in a state of rest. Its detection shows decidedly the presence of animal life in ice.

Dirt is hard to picture, but should have a place in this morphology, though it has been defined as "matter out of place."

Tabellaria.—Diatom found commonly in all surface drinking waters. They have the power to arrange in rows, and the specimen has fifteen individuals in one aggregation, which is a small one. Diatoms are regarded as plants by the majority of observers. A good deal of difficulty arises from trying to measure things with the lines and plummets of past times, when the things in question were absolutely unknown, and hence could not be properly named at the date when the word "plant" was invented. As knowledge increases, names must be changed. The diatoms are generally regarded as innocent, though some observers take the opposite ground.

Epithelia.—These are probably human, washed into

the water, and frozen into the ice. They are constantly thrown off in washing, sputa, and the excretions of the body. They are also found on all other vertebrate animals and on vegetables.

"Mycelial filaments of a vinegar yeast found in connection with melting ice. At the bottom are the embryonal spores of the yeast."—*Scientific American*, p. 73, col. 2.

This shows what happens when ice-water is allowed to stand exposed to the action of the air. A long, dirty, grayish, gelatinous ribbon, half an inch wide and about one-eighth inch thick, appeared to be a mass of what is called "the mother of vinegar." The cut gave the appearances under the microscope. The significance shows what is the full development of some of the embryonal forms of life found in ice-water when subjected to conditions that are present in refrigerators.

It must be remembered that these are not the full lists of what were examined. Some could not be named. Neither can it be said here that it has been settled that ice is injurious or not. But enough testimony is here given to indicate that ice should not be used in water; but if the water must be cooled, let it be done by placing jars of water in ice.

D. AIR.

The idea that air is food is found in Hindustanee language of three thousand years ago. The word animal infers air to sustain life. If any one doubts this position as to air being food, let him hold his breath for five minutes.

There are many ways to study this morphology, among which are:

1. Moisten the cleaned tip of one's finger with distilled or filtered water, or water whose morphology is known, then touch it to the top of some article of furniture. Instantly the tip will be covered with dust or forms that have mounted through the air to rest where found. This dust can be transferred to a slide, covered, and examined. I think this the quickest and easiest mode.

2. Ice. Let a piece of ice melt in the air to be examined. Instantly there is a current of air towards it bearing the forms against the moist surface of the ice; they stick, and can be removed on to a slide, covered, and examined under the microscope. Or, the ice may be allowed to melt in a vessel, and the resultant water explored as in water examinations.

3. Exposure of slides moistened with glycerin or not, with or without a cone attached to a vane, so that the air impinges on the slide.

4. Air may be filtered through a cotton bag, and then the bag reversed and washed in filtered water.

5. A slide may be placed on a flat surface, or on pins or legs, so as to catch the forms that fall or that are forced from below, as in ague districts.

6. Snow may be taken in a can or pail, or any receptacle that has been cleaned with dis-

tilled or filtered water. The snow allowed to melt, and the water filtered; the filtrate will be found to contain many forms.

It is astonishing how the air in motion will carry solids. In San Francisco, I saw sand from the Pacific Ocean dried and blown in such quantities as to go over houses and bury street lamps. I have read of moving mountains of sand. Perhaps the writer may say that he writes on the eighth floor of a large apartment house, where he expected to be free from the dust which annoyed him at a past residence on the second floor, but the fact is, his microscope glass table is covered in one day as much as in three at the former residence. Such facts deserve attention of those who study malaria, and such must expect to find the morphology of the air mixed with the other morphologies; still it will not do to attribute to the air things that belong to other morphologies. The carrying properties of air are underestimated by people not housekeepers.

The morphological study of the air prepares one to be careful in rejecting evidence which shows the route of invasion of diseases by the medium of the atmosphere through the air passages.

Ague plants.
Algæ.
All dusts from soils.
Anything that comes from the wear and tear

of the multitudinous operations of life everywhere, whether dried and blown by currents of wind, or by heat, or diffusion of gases.

    Asthmatos ciliaris.
    Automobile spores.
    Bi-acicular crystals, etc.
    Coal.
    Cotton.
    Crystals of chloride of ammonium.
    Diatoms.
    Epithelia.
    Fat globules.
    Feathers of birds and insects.
    Fungi spores and macrospores.
    Hairs of animals and plants.
    Insects and parts of insects.
    Leather.
    Linen fibre.
    Palmellæ.
    Paper.
    Pigment matters.
    Pollens of plants.
    Pus.
    Silica.
    Smoke products.
    Sphærotheca pyrus.
    Spores and young plants of:
        Protuberans gelatiformis.
        Protuberans lamella.
        Protuberans ovalis, with dried incrustations of the same.

Spores of cryptogamic vegetations of the sick carried by the sweat.
Starches.
Vibriones.
Volcanic dust.
Winged seeds, etc.
Woody fibre.
Wool.
Yeast spores, alcoholic, lactic acid, butyric acid, etc.
Zoospores

E. MORPHOLOGY OF FOODS, ANIMAL AND VEGETABLE.

The limits of this work having been exceeded, only a passing allusion can be made to this large, fruitful and important field, which is close at hand, easily manipulated and intensely interesting and profitable. There are four phases in which the morphology should be studied.

1. *Uncooked.*
2. *Cooked.*
3. *After migration through the alimentary canal.*
4. *Adulterations.*

1. UNCOOKED.

For example, take the potato; its skin, cortical substance and parenchyma should be studied in thin sections, and all the forms noted, whether the names are known or not. Among these are the epithelia, cork cells, connective fibrous tissues, spiral tissue, pitted ducts, gubernacula leading from the "eyes" to the centre of the parenchyma, the reticulation of a cross section, the starch grains filling such a section, as eggs in a basket, the various sizes, shapes, concentric markings of the starch grains, the action of the polarized light on the starch and cellulose, etc., etc.

2. COOKED.

By boiling, steaming, or action of hot fat. See if the starch polarizes the light; if so, the potato is not fully cooked. See the sacs of the potato substance embracing the starch grains, which, if well cooked, should be converted into a homogeneous mass all mixed up together, with no sign of the uncooked egg-shaped forms they had before cooking.

3. EXAMINED IN THE FECES

Of the eater; if any of the sacs are found, that have not been digested, the clinical examiner must study to find out if the fault lies with the alimentary canal, which has allowed the potato sacs to traverse it undigested. If the contents of the sacs are not broken up or homogeneous, and do not polarize light, the fault must lie with the digestive apparatus. Generally, when a food that is properly cooked, or raw, runs through the alimentary canal intact, it should be avoided. It is folly to give the digestive system problems which it is unable to solve. Better change to something else that will digest or administer such remedies as will make them digest. Here is a beautiful field of study; I say beautiful, because its lessons are so clear and instructive, and because some of the finest specimens of polarized light are found in the feces.

As to beef-steak.

1. Uncooked, note its beauty under polarized light, the trichinæ (if present), the physical appearances of the fibrillæ, the amount of fatty infiltration, the amount of connective tissue, etc., etc.

2. Cooked.—The shrinking in size, the absence of polarization, the darkened color approaching black.

3. In the feces, if not broken up into a fine homogeneous mass, like a solid extract in which no forms of muscular fibre can be detected, it is not thoroughly digested. If the muscular fibres are found undigested, they tell their own story plainly.

It must be remembered that the connective, areolar, and fibrous tissues from the vegetable kingdom are almost all insoluble in the juices of the alimentary canal, and must be expected to appear in the feces of healthy digestion.

The above list might be extended by including celery, cranberries, grapes, peaches, wheat, oats, barley, rye, melon, specially watermelons, which show beautifully protoplasm in active motion, tomatoes, corn, squash, sweet potatoes, mustard, bread of all kinds, cake, crackers, pilot bread, unleavened bread, wines, dough, yeast from sour bread, etc., etc., etc.

The use of the polariscope is invaluable as a test for cooking. The writer has used it for many years, and was probably the first to call attention to its great value as a test for cooking. The morphology of foods throws great light on the alcohol question.

4. ADULTERATION OF FOODS.

This department would fill a book, but attention can only be called to it here. So long as money can be made by false dealings as to foods, just so long is there need of protection by a knowledge of the morphology of adulterations of foods.

The statements of the interested parties should be tested by the microscope. For example, if an article claims to be pure coffee, it should prove to be so under the microscope. A study of a genuine grain of coffee will give the clues, and a study of chicory will also be of help, as it is generally used for adulteration of coffee. Indeed, the adulterations of all spices, black pepper, for example, with ground buttonwood bark, have been going on for years, and will probably go on till this subject is properly understood, and this will be when microscopes are as common as pianos and organs. May this time soon come!

The morphology of food is easiest of all to study, and no one should give decided opinions before practical knowledge is acquired; those who have never had their attention called to this subject, will find its investigation to be a great revelation as to human nature.

Infants' Foods.

The writer must content himself with referring to his monograph on this subject, which will be furnished on application to him. It is sufficient to say here that most of them fall short of their claims, and should be given a wide berth. Far better is it to feed during motherhood so that there shall be an abundance of healthy milk, to wit: two-thirds animal and one-third vegetable food (see "Food in Motherhood," by author, about to appear), and then there will be no need of artificial feeding of infants.

Should this present work be encouraged by

the profession, the writer will give a fuller treatment of the morphology of foods, which will involve considerable expense of time, labor, and money, and which, by good right, should not be done by private enterprise, but under governmental patronage, because it has the most intimate relations to the welfare of its most precious articles of value in the nation, to wit: the human beings within its confines.

VIII.

MORPHOLOGY OF CLOTHING.

This is a practical question, showing how to have no cheats in clothing; but it assumes a more intense interest in its medico-legal relations, for example, the examination of blood stains on coats, shirts, pockets, money bags, greenbacks, etc., etc.

Everything found in the morphology of the air and dirt must be expected here, added to the morphology of dried blood. Careful mensuration and inspection of the suspected blood must be made amid the crowd of other objects, such as silica, feathers, starch of all kinds, pollen of many kinds, pigment matters, hairs of plants and animals, fibres of textile fabrics, animal and vegetable tissues, fungi and algæ, and so forth.

Corpuscles of various shapes distorted in drying or not may be found. Now and then, perfect ones can be found alone, or buried wholly or in part in the clot.

When the stains have been washed with water to remove them, as water is the best thing for this purpose, the morphology is still more difficult. Yet making allowance for this bleaching detergent process, much valuable information can be had which, while it does not positively convict or release, points the way out to conviction or not, as the case may be, very strongly in doubtful cases. In our present state of knowledge, no one should be hung or set free simply upon the blood evidence alone, unless the claim is made that the blood stain is one of the bird family, whose corpuscles are oval and whose white corpuscles are smaller than the red. The microscope should not be made to prove more than belongs to its domain.

To examine water morphology,* filter through cotton bag, about one and one-half by four inches, with as gentle a pressure as possible. When the water begins to bore through in jets, stop flow. Remove bag, empty into a goblet, turn bag inside out and sop in goblet a short time. Squeeze bag by twisting. With a pipette remove specimens on to a slide and cover, or, better, have a slide with an open cell, two by two-thirds inch, one-eighth inch deep, and place specimen on horizontal stage; one inch, one-quarter to one-tenth inch objectives.

---

* See pages 49 and 58.

www.ingramcontent.com/pod-product-compliance
Lightning Source LLC
Chambersburg PA
CBHW031120160426
43192CB00008B/1054